D. L. Mundy

Rotomahana and the Boiling Springs of New Zealand

A Photographic Series of Sixteen Views

.

D. L. Mundy

Rotomahana and the Boiling Springs of New Zealand
A Photographic Series of Sixteen Views

ISBN/EAN: 9783337373061

Printed in Europe, USA, Canada, Australia, Japan

Cover: Foto ©Andreas Hilbeck / pixelio.de

More available books at **www.hansebooks.com**

ROTOMAHANA;

AND

THE BOILING SPRINGS

OF

NEW ZEALAND.

A PHOTOGRAPHIC SERIES OF SIXTEEN VIEWS

BY

D. L. MUNDY.

WITH DESCRIPTIVE NOTES

BY

FERDINAND VON HOCHSTETTER,

PROFESSOR OF THE POLYTECHNIC INSTITUTION OF VIENNA.

"O ye Fire and Heat, bless ye the Lord;
Praise him, and magnify him for ever!"

LONDON:
SAMPSON LOW, MARSTON, LOW, AND SEARLE,
CROWN BUILDINGS, FLEET STREET.
1875.

TO THE

HONOURABLE JULIUS VOGEL,

PRIME MINISTER OF NEW ZEALAND,

THIS ACCOUNT OF THE SOUTHERN WONDERLAND, ROTOMAHANA,

AND THE BOILING SPRINGS OF NEW ZEALAND,

IS DEDICATED

BY

D. L. MUNDY.

CONTENTS.

I.

The Photographs are reproduced for this publication by the Autotype Process, which renders the Pictures permanent.

THE HOT SPRINGS DISTRICT OF NEW ZEALAND.

THE official Government papers of New Zealand contain a letter from the Hon. W. Fox, of Wellington, to the Prime Minister (now Sir Julius Vogel, K. C. M. G.), dated August 1, 1874, in which he recommends that the Colonial Government should take steps for the acquisition of the Hot Springs District, as a public domain, with a view to utilizing this wonderful provision of nature for sanitary purposes. Mr. Fox, after minutely describing the lakes and springs, goes on to observe :—

" It is not my intention to dilate on the wonderful and beautiful which abound in connection with Rotomahana and its terraces. A day spent among them is a new sensation, and must be felt to be understood. I wish rather to draw attention to the different groups of springs, with a view to their sanitary use. At the same time, the idea that these majestic scenes may one day be desecrated by all the constituents of a common watering-place, has something in it bordering on profanity. I would not suggest that their healing waters should be withheld from the weary invalid or feeble valetudinarian. Doubtless their sanitary properties were given them for the good of suffering humanity, and that they should become the Bethesda of New Zealand would detract nothing from the sanctity and grandeur. But that they should be surrounded with pretentious hotels and scarcely less offensive tea-gardens ; that they should be strewed with orange-peel, with walnut shells, and the capsules of bitter-beer bottles (as the Great Pyramid and even the summit of Mount Sinai are), is a consummation from the very idea of which the soul of every lover of nature must recoil. The Government of the United States had hardly become acquainted with the fact that they possessed a territory comprising similar volcanic wonders at the forks of the Yellow River and Missouri, than an Act of Congress was passed reserving a block of land of sixty miles square, within which the geysers and hot springs are, as public parks, to be for ever under the protection of the States ; and it will doubtless take care that they shall not become the prey of private speculators, or of men to whom a few dollars may present more charms than all the finest works of creation.

" I beg to suggest to the Government of New Zealand that as soon as the Native title may be extinguished, some such step should be taken with regard to Rotomahana, its terraces, and other volcanic wonders. It is to the credit of the Maoris that they have hitherto done all in their power to protect them, and express no measured indignation at the sacrilegious act of some European barbarians who, impelled by scientific zeal or vulgar curiosity, have chipped off several handsbreadths of the lovely salmon-coloured surface of the Pink Terrace.

" I have endeavoured in this imperfect sketch which I have given (and for the details of which I am much indebted to Hochstetter, correcting my own less careful observation,) to draw the attention of the Government to the great value of the sanitary provision which nature has made in the district described. I think the time has come when something practical might be done to utilize that provision. At present, the difficulty of travelling in the Hot Spring country, and the almost entire absence of accommodation for invalids, prevents more than a very small number of persons from visiting it, either for health, recreation, or curiosity. Yet it might be, and is probably destined to be, the sanatorium not only of the Australian Colonies, but of India and other portions of the globe.

" I have, &c.,
" WILLIAM Fox."

" The Hon. the Premier."

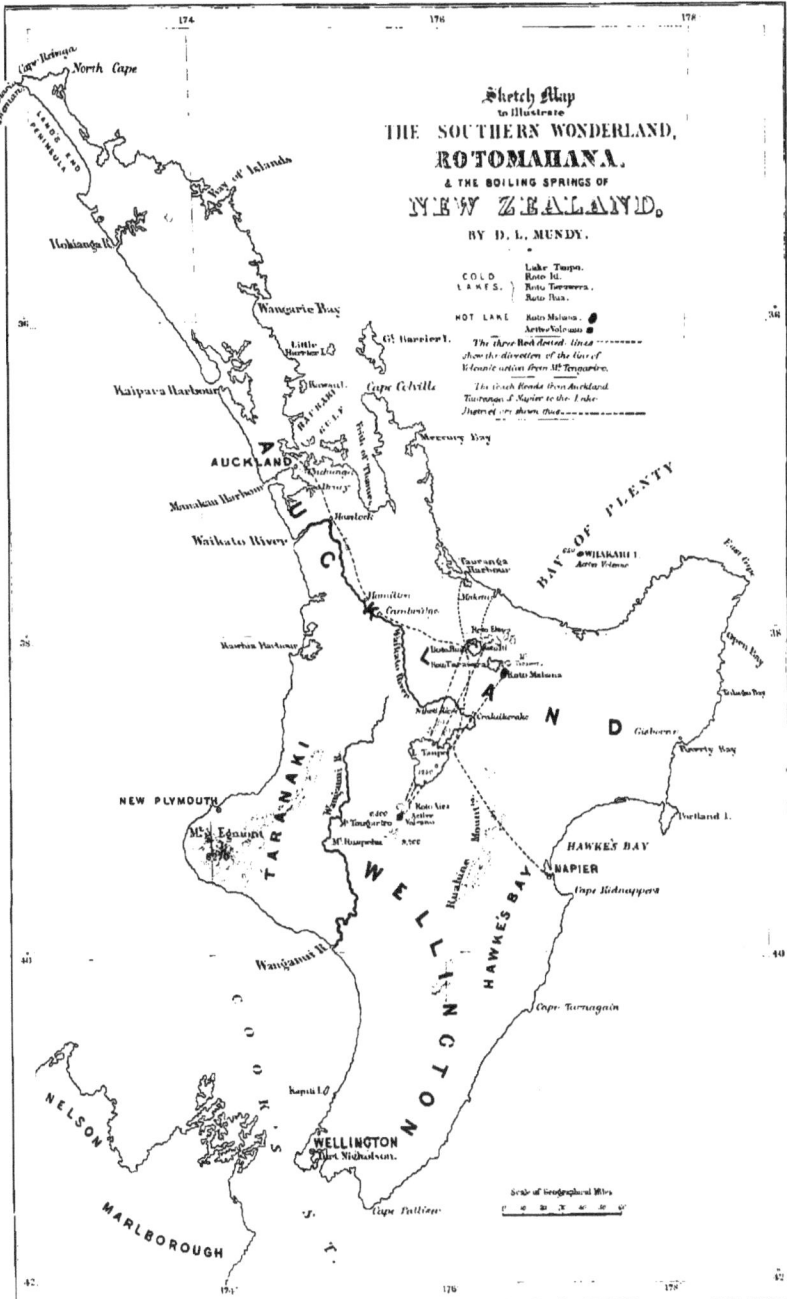

Sketch Map
to Illustrate
THE SOUTHERN WONDERLAND,
ROTOMAHANA,
& THE BOILING SPRINGS OF
NEW ZEALAND.
BY D. L. MUNDY.

COLD LAKES.
Lake Taupo.
Roto Id.
Roto Tarawera.
Roto Rua.

HOT LAKE Roto Mahana.
Active Volcano

The three Red dotted lines
show the direction of the lines of
Volcanic action from Mt Tongariro.

The Coach Roads from Auckland
Tauranga & Napier to the Lake
District are shown thus

North Cape
Cape Maria
Cape Reinga
REINGA PENINSULA
Hokianga R.
Whangarie Bay
Bay of Islands
Kaipara Harbour
Little Barrier I.
Gt Barrier I.
Cape Colville
Mercury Bay
AUCKLAND
Manukau Harbour
Waikato River
Huntock
Raglan Harbour
Tauranga Harbour
BAY OF PLENTY
WHAKARI I
Active Volcano
East Cape
Hamilton
Cambridge
Mokau
Open Bay
New Plymouth
Mt Egmont
TARANAKI
Taupo
WELLINGTON
Wanganui R.
Rotorua
Roto Tarawera
Roto Mahana
Orakeikorako
Gisborne
Poverty Bay
Mt Tongariro
Active Volcano
Mt Ruapehu
HAWKE'S BAY
NAPIER
Cape Kidnappers
Portland I.
Cape Turnagain
Kapiti I.O
WELLINGTON
Port Nicholson.
Cape Palliser
NELSON
MARLBOROUGH
COOK'S STRAIT

Scale of Geographical Miles

London: Sampson Low, Marston, Low & Searle, 188, Fleet St.

THE HOT SPRINGS DISTRICT OF
NEW ZEALAND.

HE two large islands which, together with one of insignificant size, constitute New Zealand, perhaps the most attractive and most flourishing of British Colonies, differ considerably in their natural aspects. Both are traversed lengthwise by great mountain ranges, from south-west to north-east, which would form one continuous range but for Cook's Strait between the two islands. The mountain scenery, indeed, of the South or Middle Island, has an Alpine character, with its lofty snow-covered peaks, its immense glaciers, its profound and precipitous ravines, and beds of shattered rock. On the west side of this mighty range, the sea-coast is overhung with dense dark forests; on the east side lie the open plains of Canterbury and of Otago. In the North Island, on the contrary, the upheaved mass of stratified rocks forming the backbone of the country, while it still pursues the same general line of direction, approaches the eastern shore, declining to half the height of the Southern Alps. A broad tableland is left in the middle width of this island, to the west of the backbone range, and is buttressed near the western sea-coast by Mount Egmont. This ground, from Cook's Strait to the Bay of Plenty, is chiefly of volcanic formation, consisting of pumice-stone, tufa, trachyte and rhyolite lava, at first formed by submarine eruptions, but afterwards gradually lifted above the sea. Legends still current among the Maoris, or native people, who call their land Te Ika a Maui, " The Fish of Maui," would seem to be derived from a mythical tradition of its rising from the ocean. This is only one of three distinct volcanic zones in the North Island; and the other two, which lie across the narrow northern peninsula, at Auckland and at the Bay of Islands, are of more recent origin; but their volcanic action is extinct. The region above mentioned, within which is comprised the Hot Springs district, exhibits some of the grandest and rarest volcanic phenomena to be seen on earth. It will suffice, in this introductory chapter, to give the mere outline of a general description.

From the opposite south-west and north-east shores of this territory—that is to say, from Cook's Strait on the one hand, and from the Bay of Plenty on the other, the land gradually rises to an elevation of 2,000 feet. Here stand two high mountains, Ruapahu and Tongariro; the former being an extinct volcano, 9,200 feet in height; the latter, about 6,500 feet, one that is still active. Away to the west, upon a parallel but different line, stands Taranaki or Mount Egmont, 8,270 feet, a third volcano, but, like Ruapahu, now extinct. From the yet burning crater of Tongariro, in a straight line drawn to the north-east for 200 miles as far as Whakari or White Island, an active volcano twenty-eight miles from the shore in the Bay of Plenty, extends the present range of igneous forces beneath the surface of land and sea. The Maori legend may here again be cited, which relates that a celestial personage, Ngatiroirangi, or the "Runner of the Sky," having alighted upon the snowy summit of Tongariro, felt himself suffering from the cold; he shouted to his sisters, who remained on Whakari, to send him some fire which they had brought from Hawaiki, the mystic home of gods and men. They did so, by the agency of the two Taniwhas, the Earth Spirit and the Water Spirit, who dwell underground. The fire thus transmitted is still current through the long subterranean passage, and bursts forth in a thousand places, from the soil of the plain, the side of the mountain, or the boiling depths of streams and lakes. Another story is, that the national demigod Maui, when he saw the mighty "Fish," as this land is called, emerge out of the sea, was alarmed by its strange fires, and kicked or tossed them along from Tongariro to the Bay of Plenty, in order to quench them. These fables, as poetical as those of the ancient Greeks which professed to account for marvellous physical phenomena, are based also upon correct observation of the line of volcanic action in New Zealand.

Of the wonderful and beautiful effects of such action, as faithfully illustrated by the series of Sixteen Photographic Views presented in this publication, the reader will presently judge for himself. They are similar in kind to others long since renowned, such as the geysers, the solfataras, the steam jets, and other remarkable features of Iceland; as well as the grander display of these natural miracles, lately discovered and examined by a scientific expedition in America, at the head waters of the Yellowstone River. But the more characteristic and peculiar features of the New Zealand Hot Springs region, such as the Cascade Terraces of Rotomahana, have no rival in any part of the globe. The system to which they belong, regarded in general, surpasses every other in variety and extent. We must remark, too, the manifest unity which it possesses, from Lake Taupo, a stupendous central reservoir of water at a great elevation, down to the Rotorua, the Rotoiti, and the Rotoehu, near the sea coast of the Bay of Plenty. To comprehend this system as a whole, throughout the survey of many diverse effects of its subterranean volcanic forces in the disturbance of land and water, it is needful that we should first view Lake Taupo as its actual head. The same lake has another bearing of great importance on the physical geography of the island, from its connexion with the large river Waikato. Few things, indeed, are more singular than the position of Lake Taupo itself, in relation to the general structure of the country where it holds a central place.

The Maori name of Taupo, which means "Darkness," seems to express the sense of mystery that ever invests, to their superstitious imagination, this bottomless Moana, or inland sea, around which yet linger the remnants of the nation, preserving their genuine customs and manner of life. In the centre of the land, 160 miles from the coast of the Bay of Plenty, the lake fills a cavity of unknown depth, a huge hole in the midst of the elevated plateau of volcanic formation. Its water stands at the level of 1250 feet above the sea, yet no soundings have been reached at 200 fathoms. The lake may probably, in former ages, have stood several hundred feet higher, as the Maoris assert that it did; for its shores exhibit three successive terraces or plains, the lower being alluvial, which rise above one another to the general level of the surrounding table-land, 700 feet or 800 feet above the present water. Its area is about 248 square miles, and the length of its coasts nearly 150 miles. The extreme length, N.N.E. to S.S.W., from Tapueharuru to Waihi, is 25 statute miles; the breadth, from Hamaria, on the east side, to Maungahara, on the west, is 20 miles. At its southern extremity,

Lake Taupo receives, in the delta about Tokanu, the four branches of the upper Waikato, having their sources at the feet of Tongariro and Ruapahu and neighbouring mountains. Several other rivers enter the lake on its eastern side. But its only visible outlet is the exit of the Waikato at the north end of the lake, through which that river flows like the Rhine flowing through the Lake of Constance, or the Rhône through the Lake of Geneva. It may, however, be conjectured that a large portion of the water of this vast reservoir is carried off northward by subterranean channels, reappearing in the innumerable hot springs and the many smaller lakes between Taupo and the Bay of Plenty coast. The actual volume of water discharged by the Waikato, as well as the dimensions of the lake, were ascertained at my request by Mr. C. Maling, the government surveyor. It is computed to be 245½ millions of gallons per hour during the lowest summer level of the lake; but in the winter months, July, August, and September, it amounts to 280 million gallons.

The volcanic rocks around Lake Taupo consist of quartzous trachytic lavas, now called rhyolite, in a variety of modifications, crystalline and vitreous, such as obsidian. Together with masses of pumice-stone, they form cliffs on the western shore, to the height of 1,000 feet and more at the promontory of Karangahape; and beyond those tall bluffs rise, in the distance, the wooded ranges of Rangitoto and Tahua, 3,000 feet high, with the pyramidal summit of Titiraupenga. The eastern shore, on the contrary, is mostly flat, with a broad beach of pumice stone, sand, or gravel, across which the Waimarino, the Tauranga, the Hinemau, and the Waitanui flow into the lake. Its most striking features are the rocky peninsula Motuoapa, the wooded island of Motutaiko, with its precipitous northward front, and the steep banks near Motutere. But still more worthy of inspection are the cliffs at Totara, which are curiously variegated, and of a very regular columnar formation, consisting of thin laminated sheets of various rocks, fused together in a streaky mass. Farther north, the visitor's attention is caught by the deep channels which rivers have cut in the pumice-stone tableland above. The Tauhara mountain, an extinct volcano, lifts its graceful head above the outlet of the Waikato. It is, however, by looking southwards up Lake Taupo, that a prospect is gained of the crowning eminences of this region. To the left hand, beyond the wooded range of Kaimanawa, which is ten or fifteen miles east of the lake, are the towering Alpine peaks of the backbone mountains of New Zealand, so different in their aspect from the volcanic cones; these mountains continuing the Ruahine chain from the south, but assuming as they become lower the name of Te Whaiti. But immediately before the spectator, at the south end of the lake (see View XIII.) the giant forms of Tongariro, breathing forth a cloud of steam from its principal crater, and of Ruapahu, clad with perpetual snows, arise behind a group of lesser volcanic mountains. These are Pihanga, called the wife of Tongariro, with Kakaramea, Kuhaua, and one or two others, which the Maoris have fancied to be Tongariro's children. They say that Taranaki was kicked by Tongariro as far off as New Plymouth for his insolence in making love to Pihanga. There is no record of any eruption of lava from the still active craters of Tongariro, the chief of which, called Ngauruhoe, is at the top of a cinder-cone surrounded by a circular wall of cliff, 1,000 feet high, with but one opening, to the west. Cinders and ashes, and hot mud, have been thrown up now and then, breaking the lip of the crater; but its ordinary discharges consist of steam, or sulphurous vapour. It stands about twelve miles from the south end of Lake Taupo.

Such as we have described is the head of the Hot Springs region; and here, in the neighbourhood of Tokanu, at the southern or upper end of Lake Taupo, are found numerous places where the subterranean forces break through the ground, as shown in the Views. But from this locality to the sea-coast, in a direction 36 degrees E. of N., three parallel lines of similar volcanic action, not far apart, may be distinctly traced. One is the line drawn from Tongariro to the marine volcano Whakari; and this line includes the hot springs at the head of Lake Taupo, and those of Rotomahana, a long way farther on beyond the Kaingaroa plain. The second line, which follows, to some length, the outflow of the river Waikato from Lake Taupo, is marked by the hot springs and steam jets of Otumaheke and Orakeikorako, on the river's banks, and those of the Pairoa mountain range. The third line of action forming eruptions of this kind is exhibited in the hot springs of Rotorua and the solfataras of Rotoiti, which terminate the notable specimens of volcanic phenomena on land, being situated near the sea coast. It is at Rotomahana, more especially, that the most beautiful and marvellous effects of this action are displayed.

The cause and mode of their production may be explained with brevity in general terms. Water, having descended through fissures into the earth, is heated by volcanic fires, and generates high-pressure steam. This rises, accompanied by sulphurous acid and other gases of volcanic origin, which decompose the nether surface of the lava or other rock that overspreads the neighbouring region. The steam is, meanwhile, partly condensed to hot water; and this, impregnated with acid gases and soluble substances, forms, at the surface, the springs. All the New Zealand hot springs, like those of Iceland, are to be divided into two distinct classes, the one alkaline, the other acid. To the latter belong all the sulphurous hot springs on the Rotomahana, Rotorua, and Rotoiti, characterized by deposits of sulphur or frequent incrustations of alum, which having no periodical eruptions, are either tranquil or in a uniform state of ebullition. This first class of thermal springs is called by the Maoris, Ngawhas, corresponding with those in Iceland called Namur. The second class of alkaline springs is that which bears in Iceland the designation of Hverjar, and to which belong the Geyser and Strokkur; but in New Zealand these intermittent fountains are called Puias, or bursting craters. This word is especially used in the Taupo country to designate the intermittent, geyser-like fountains of Tokanu, of Orakeikorako on the Waikato, and of Whakarewarewa on Lake Rotorua. All the intermittent springs belong to the alkaline class, which, on the cooling and evaporation of the water, deposit great quantities of silica. Springs suited to bathing purposes, the water of which never reaches the boiling point, and all naturally warm baths, are called Waiariki, corresponding to the Laugar of Iceland.

I am indebted to Professor Ferdinand von Hochstetter for much of the information contained in this general account of the Hot Springs district of New Zealand, and he has also kindly furnished the Descriptive Notes, which follow, upon the particular subjects of my Photographic Views.

D. L. MUNDY.

LONDON, *March 11th, 1875.*

I.

THE ROTOMAHANA, OR HOT LAKE.

NORTH-EAST VIEW, WITH THE TARAWERA MOUNTAIN.

THE Rotomahana is one of the smallest lakes, being scarcely a mile in length from north to south, and only a quarter of a mile in breadth. Yet it is the most marvellous of all, being fed by boiling springs, which keep the whole volume of its water at a high temperature. They deposit an enamel-like white siliceous crust or coating over the margin, which gives to the entire region a fantastic and Fairyland-like appearance. The Rotomahana is situated 1,088 feet above the level of the sea.

The shape of the lake is very irregular. On the southern side, where the shore is formed by swamps overgrown with bulrushes, warm water gushes forth; hot mud-pools are seen here and there; small rivulets flow into the lake; and from projecting points muddy shallows, covered with swamp-grass, extend almost to the middle of the lake. At its northern end the lake narrows, and, where the Kaiwaka rivulet flows out, there are again grass-swamps and shallows. In the middle, however, the water is deeper, and the shores to the east and to the west are steep and rocky. The Rotomahana is correctly named " The Hot Lake ;" for the quantity of boiling water issuing from underground, both on the shores and at the bottom of the lake, is truly astonishing. The temperature, however, is not uniform, but differs in various places. Wherever the rising of gas-bubbles indicates a hot spring at the bottom of the lake, the thermometer will frequently rise to 90° or 100° Fahr. Near the mouths of the cold rivulets, the water of which shows a temperature of 50° to 52° Fahr., the temperature of the lake water has been found to be only from 60° to 70° Fahr. But in the middle of the lake, and near its outlet, the Kaiwaka channel, at the north-east end, it rises to 80° Fahr., which may be considered the mean temperature of the lake. In bathing or swimming in different parts of the lake, the variation of temperature is easily noticed, and care must be taken to avoid approaching too near to any of the hot springs. The water is muddy and of a turbid green colour. Neither fish nor mussels live in it; on the other hand, this lake is a favourite haunt of countless water-fowl and morass-fowl. Various kinds of ducks, water hens, the magnificent *Porphyrio melanotus*, and the graceful oyster-catcher, *Hæmatopus picatus*, frequent Rotomahana, and make these warm shores their breeding-place, while they resort for their food to the neighbouring cold-water lakes.

The main interest attaches to the eastern shore. Here are the principal springs to which this lake owes its fame. They are shown in the accompanying view, ranging in the direction from north-east to south-west, in the following order, beginning on the left-hand side of the view : Te Tarata, Ngahapu, Te Takapo, the Steaming Ranges, and the Rua Kiwi, with the two islands Pukura and Puai. The foreground of this view is the foot of the Pink Terrace, called Otukapuarangi, on the south-west side of the lake ; in front of which, on a promontory in the middle of the view, is the great sulphur lake or solfatara, called Te Whakataratara.

First, there is the Te Tarata, or the White Terrace, at the north-east end of the lake, with terraced marble steps projecting into the water. This is the most remarkable of the Rotomahana wonders. From the foot of the Te Tarata a path leads by the slope of the hill to the Great Ngahapu or Ohapu. This fountain, shut in by a thicket of shrubs, lies close to the margin of the shore, about ten feet above the level of the lake. The huge columns of steam continually ascending from it show its locality from afar. Its basin is forty feet long by thirty feet wide ; the water is clear and transparent, but almost always in a terrific state of ebullition. The seething foam rises to a height of eight or ten feet, and a surf of hot waves lashes the walls of the basin. The elevated margin of siliceous deposit prevents the free overflow of the water ; the natives have, therefore, constructed an artificial outlet, conducting the water to several bathing basins. The thermometer rises in the hot spring to 210° Fahr. The siliceous deposits are of a dirty brown colour. Farther up on the hill lies the Little Ngahapu, a basin in which turbid, muddy water seethes ; but this is difficult of access.

Near to the Great Ngahapu, but more to the south and close to the shore, is Te Takapo, an incrusted basin, ten feet long by eight feet wide, with clear, gently-boiling water of 206° Fahr. This geyser is said to rise occasionally in a jet from thirty to forty feet high. There are numerous smaller springs, bubbling mud pools, and lightly incrusted holes along the shore, between Te Tarata and Te Takapo ; several deserted huts stand here. A few yards further is a ravine, called Waikanapanapa, or Boiling Mud Lake, extending in the north-east direction a quarter of a mile. Some distance back, and about a hundred feet above the level of the Rotomahana, lies the Rotopounamu, or Green Lake, a dirty green-looking water-basin about forty feet in diameter. Southwards, at the mouth of the ravine, lies the fountain Rua Kiwi (Kiwi Hole). From the Waikanapanapa valley, opposite the two islands Pukura and Puai, the shores of the lake become steep and rocky ; hot springs gush out from below, under the surface of the lake ; while above, on the side of the hill, are the huts of Ngawhana or Ohana, scattered near the spring which bears that name.

To the Ngawhana group belong several hot springs which have all contributed to the siliceous deposits covering the slope of the hills. The Ngawhana itself is a basin of hot but still water. Farther up, about a hundred feet above the lake, is the Koingo (sighing), so called from the sighing sound echoed by the water replunging into the basin. This is an intermittent fountain, from which discharges of water take place three or four times a day, alternating—as the natives say—with those of the neighbouring Whatapoho. This Whatapoho is one of the most remarkable springs on the borders of Rotomahana Lake— being partly a fountain, partly solfatara, and partly fumarole, or, strictly speaking, all three in one. From a deep shaft-like aperture between brittle, ash-coloured rocks, ascend, as from a steam boiler, volumes of hot steam and sulphurous gas, with a dismal, moaning sound. It is dangerous to approach the basin too closely.

II

THE ROTOMAHANA.

THE south side of the lake does not present a single spring of any size, worthy of note; but on the western shore, on the right-hand side of the view, is the great Pink Terrace Cascade, Otukapuarangi, or " Cloudy Sky," which takes that name from the immense cloud of steam always rising above it. This is a counterpart of the Te Tarata cascade. On the left-hand side of the view in the middle ground appear the islands Pukura and Puai. Altogether about twenty-five large Ngawhas (the native name for hot springs) may be counted on the Rotomahana. The smaller springs, which bubble up at innumerable places over an area of about two square miles, are beyond counting.

It is gratifying to learn that these hot springs, according to the experience of the natives and European visitors, have proved, very effective in curing chronic cutaneous diseases and rheumatic pains. There is little doubt but at no distant period the Rotomahana, with its genial climate, its wonderful cascades, hot medicinal springs, and ready-made warm baths, will become, not only an attractive resort for tourists, but also the sanatorium of Australia, and a place of cure for invalids from all parts of the world.

III.

TE TARATA, OR THE WHITE
TERRACE CASCADE.

FRONT VIEW.

THE name of Te Tarata, or tattooed rock, is derived from the peculiar forms or figures presented by the siliceous deposits and stalactites on the terraces over which the water descends in front of the geyser crater. This cascade is situated at the north-eastern extremity of the Rotomahana lake, and forms one of the most remarkable natural phenomena in existence. At an elevation of about 100 feet above the level of the lake, on the fern-covered slope of a hill, from which hot vapours ascend in various places, lies an immense crater-shaped boiling cauldron, with steep side-walls of a glaring red colour from thirty to forty feet in height, and open only on its western side towards the lake. The size of the basin is about eighty feet in length and sixty in width. It is filled to the brim with perfectly clear and transparent boiling water, of a beautiful turquoise-blue colour. At the margin of the basin the temperature of the water was found to be 183° Fahr.; but in the centre, where the water is in a state of constant ebullition, rising with a foaming crest several feet high, it probably reaches seething point. Thick clouds of steam curl up, generally obstructing in part the view of the surface of the water; but the hissing noise of boiling and seething is always distinctly audible.

IV

TE TARATA, MOUTH OF THE BOILING GEYSER.

R. MUNDY, during his visit to Rotomahana, had the good fortune to be enabled to see the crater empty, and to take a photograph of it. He had been told, by the Maori chief Parakia, who accompanied him, that few natives had ever seen the crater empty. This, in fact, takes place but rarely, and there is a notion that it occurs only during violent easterly gales. The whole mass of water is then suddenly thrown out with immense force. On such occasions the empty basin is open to sight to a depth of about thirty feet; but it fills again very quickly. Mr. Mundy was just in time, before the boiling water rose again in the frightful aperture of this geyser, to secure the accompanying view of it. It was with difficulty that his attendant and interpreter could be persuaded to descend and figure in the representation. Ten minutes after the photograph was taken, the geyser, with a dull hissing noise, filled up its whole crater. The violent eruptions of this geyser, at long intervals, can only be compared in grandeur to those of the famous Great Geyser in Iceland. The Te Tarata basin, however, is larger than the Icelandic geyser basin, and the mass of water thrown out therefore must be immense.

V

TE TARATA, TERRACES AND CASCADE.

FROM THE EDGE OF THE CRATER.

T HE formation of these terraces is very remarkable. The water possesses in a high degree petrifying or rather incrusting qualities. The deposit is like that of the Iceland geysers, siliceous, not calcareous.

By the accumulation of these siliceous deposits and incrustations, a series of terraces has been formed on the slope of the hill; these terraces being white, as if cut from marble, present a spectacle of superb magnificence. The overflowing water looks like a cataract plunging over natural steps, and in its fall suddenly transformed to stone. To realize the full effect of this strange and wonderful display of nature's creative power, one must have climbed those steps, and studied the details of their marvellous structure.

The siliceous deposits cover an area of about three acres, and in the formation of these terraces such as they are seen at the present day, thousands of years have doubtless left the traces of their unremitting work.

The level part at the foot of the terraces extends far into Rotomahana Lake. They commence from that level with low steps, containing shallow water-basins. Farther up, the terraces increase in height from two or three to four and even eight feet. They are formed by a number of semicircular stages, no two of which are of equal height; each of these stages has a small raised margin, from which slender stalactites hang down upon the lower stage. Each platform contains one or more basins, sometimes five or six, resplendent with beautiful blue water. These form natural baths, such as no human art could have constructed, or fitted in a more luxurious and commodious style.

VI

TE TARATA, SIDE VIEW, SHOWING THE
HOT BATHS.

HE natural bathing basins of Te Tarata, as here represented, are of every variety of size and form, and differ in temperature; they may be chosen large or small, shallow or deep, or of any degree of warmth: on one terrace the visitor may enjoy a tepid bath; on another, one of a higher temperature, and so on to almost boiling heat. The basins upon the higher stages, nearer the crater, contain hotter water, of course, than those upon the lower terraces; some of the basins are so large and deep, that several persons at once can easily swim in them. The bather can have his choice; he must take care, however, to do so under the guidance of a Maori.

In ascending the steps of the terraces of the Te Tarata, one has, of course, to walk round the basins and step through the tepid water that flows over their edges upon the platform, but this is rarely above the ankle. During violent emissions of water from the crater, steaming cascades are pouring down. At ordinary times but very little water ripples over the terraces, and only the discharges on the southern side form a hot steaming waterfall.

Beyond the highest terrace is an extensive platform with a number of basins five or six feet deep; the water in them is of a temperature of from $90°$ to $100°$ Fahr. In the middle of this platform, close to the brink of the crater, is a sort of island rock, about 12 feet high, overgrown with manuka scrub, mosses, lycopodium, fern and other vegetation. This can be reached sometimes without danger; and from its summit the traveller has a fair and full view into the boiling and steaming cauldron.

Such is the famous Te Tarata, which is quite a unique object on the face of the earth. The pure white of the siliceous deposit, in contrast with the blue colour of the water, seen when the steam blows off, and the green reflection of the surrounding foliage and the intense red glare of the bare earth walls of the crater, surrounded by the whirling clouds of steam, afford a scene of unequalled grandeur and beauty. The scientific observer too may here collect an abundance of exquisite specimens of delicate stalactites, of incrusted branches, leaves, and other matters, for whatever lies upon the terraces becomes incrusted in a very short time.

VII

WAIKANAPANAPA OR THE BOILING MUD LAKE.

A S already mentioned in the description of Rotomahana, the Waikanapanapa, in the neighbourhood of Te Takapo, is a ravine on the eastern shore of the lake, extending in a north-easterly direction for a quarter of a mile.

The entrance to this ravine, overgrown with brushwood and manuka trees, is rather difficult of access. It also requires great caution to pass through it, because of the many unsafe places, where the traveller runs danger of being swallowed up in hot mud. This may even be the fate of a dog, as Sir G. Bowen told Mr. Mundy happened to a valuable retriever of his own, which here got off the track. But the Maoris have laid brushwood for the traveller to step upon, and he must be careful in walking behind his native guide alongside the morass. The inside of the ravine is much fissured and torn; odd-looking rocky serratures, threatening every moment to break loose, loom up like dismal spectres from the red, white, and blue clay, the last remains, evidently, of decomposed rocks. The bottom of the ravine consists chiefly of mud in a fluid state, with thick, flat fragments of layers of siliceous deposit lying scattered about, like cakes of floating ice after a thaw. In one place, a big cauldron of mud is seen simmering; in another, a deep basin of boiling water is bubbling; next to this gapes a terrible hole, emitting hissing jets of steam; and farther on are numerous small mud cones, from two to five feet in height, some of which, like miniature volcanoes, throw up hot mud from their craters, with a deadened rumbling sound. Others send forth volumes of sulphurous gases. On looking into their apertures, beautiful crystals of sulphur are seen hanging from their edges.

VIII.

ROTOPOUNAMU, OR THE GREEN LAKE, WITH THE STEAMING RANGES.

SOME distance beyond the boiling mud-pool at the head of the ravine already described, and 100 feet above the level of Rotomahana, lies the Green Lake, or Rotopounamu, which is also called Rotomakariri, or Cold Lake, on account of its temperature, being only 60° Fahr. This circumstance is remarkable, since it has in its vicinity boiling springs of water, solfataras and fumaroles ; and from the surrounding ranges, composed of pipeclay and rocks of various colours, innumerable jets of steam are constantly discharged, with a terrific hissing noise ; yet on these steaming ranges, around and beside the jets of hot vapour, the manuka scrub flourishes with great luxuriance.

The water basin of Rotopounamu is forty feet in diameter, and of an oval shape. The water itself appears in the lake of a dirty green colour, which, being opaque, reflects, as in a mirror, the wonderful spectacle of the steaming ranges. But any portion of this water taken up from the lake is perfectly clear, and not unpleasant in taste, being slightly acid. Rotopounamu is probably an extinct spring. This lake is surrounded by flat detached slabs of siliceous deposit, several inches thick, in every conceivable position : which seem to be fragments of the broken terrace of a former geyser now extinct.

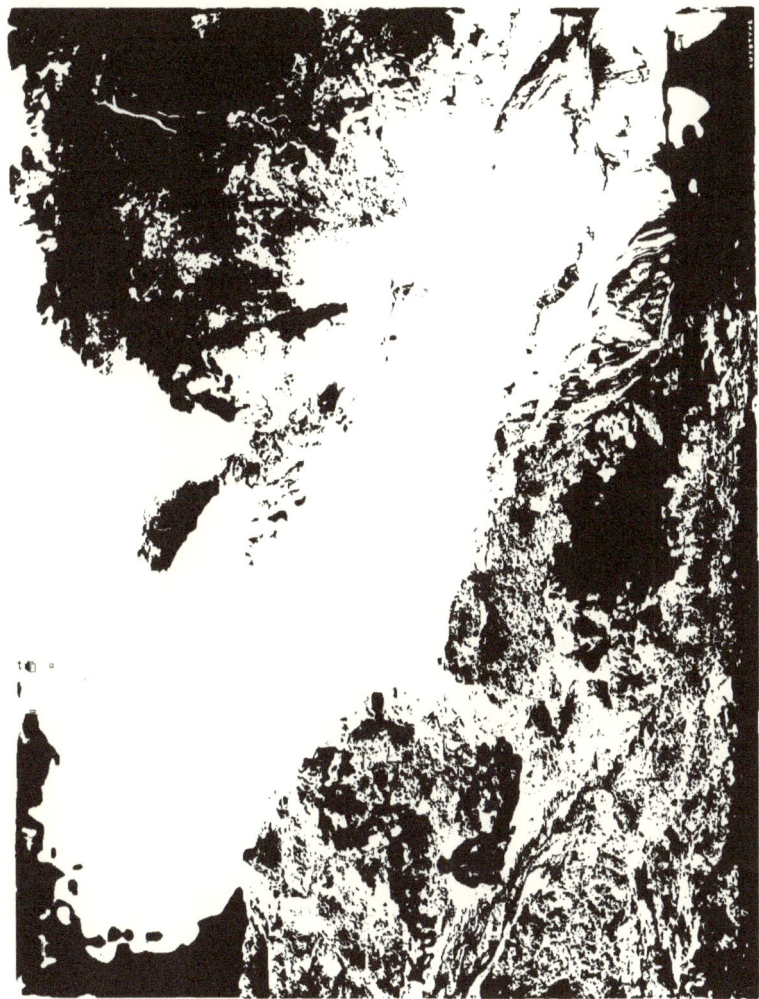

IX

RUA KIWI, OR THE BOILING CASCADE GEYSER.

T the foot of the Waikanapanapa ravine, and on the shore of Rotomahana, amidst rocks and bushes, about forty feet above the level of the lake, is the Rua Kiwi or Kiwi-Hole. This is a basin sixteen feet long and twelve wide, with clear water of 210° Fahr., which is in a constant state of gentle ebullition. The hot water continually overflows and pours down the slope, forming a perfect cascade, which is every now and then hidden by its own steam.

The stony deposit around the crater is of a dirty brown colour, instead of being white, as in the Te Tarata and other geysers. It extends as far down as the edge of the lake, where there is another and smaller hot spring, Te Kapiti.

Mr. Mundy was told by his native guide that a sad accident happened here a few years ago; a little girl, carrying a baby on her back, was looking into the Kiwi boiling geyser, when she overbalanced herself and fell into the cauldron. Death must have been instantaneous.

X

OTUKAPUARANGI, OR THE PINK TERRACE CASCADE.

ARVELLOUS as is the Te Tarata fountain with its wonderfully constructed marble-like terraces and natural bathing basins, at the north-eastern end of the Rotomahana Lake, not less resplendent in natural beauties is its companion on the western shore, the great terrace-fountain Otukapuarangi (*i.e.* Cloudy Atmosphere), so called from the continually ascending steam-clouds. Its terraces, formed of a peculiar pink-coloured siliceous deposit, extend by gentle descents, one below the other, down to the edge of the lake. On each side, the whole of the adjoining hill is thickly overgrown with manuka, manuwai, and tumingi bushes. The terraces, though not so grand and imposing as at the Te Tarata, are much neater and finer in their structure. Their enamel-like substance is tinted with the most delicate hues, varying from a light flush to a bright pink and orange chrome or salmon-colour, running in streaks down the entire depth of the terraces. This marvellous formation has the charm of a scene in Fairyland. It resembles a lovely bank of variegated coral; and, when it is lighted up by the sun, with the clear water running over it, the spectator feels transported with amazement and delight.

The crater platform is at a height of about 60 feet above the level of the lake; it is 100 yards long, and equally broad. The successive terraces or ledges, unlike those of Te Tarata, contain no basins of hot water. But on the upper platform are several basins, three or four feet deep, full of transparent sky-blue water of 90° Fahr. or more, forming delicious baths.

It is the luxury of luxuries to sit here upon a bed of soft powdery substance, looking round upon the wonderful scenery of Rotomahana, and listening to the cry of the wild birds, on a clear moonlight night. In the background, shut in by partly bare walls, tinged with various colours, red, white, and yellow, lies the large basin of the spring, like a great cauldron, forty or fifty feet in diameter, and apparently very deep. The water is usually quiet, and merely simmering, not boiling; its temperature reaches 179° Fahr. It is of an intense opal-blue colour. The ascending vapours strongly smell of sulphurous acid. The platform, for ten or fifteen feet around the edge of the crater, is covered with a thin deposit of sulphur. Here and there on the walls of the crater thick crusts of sulphur have been deposited by the ascending vapours.

XI

OTUKAPUARANGI, SIDE VIEW OF ONE OF THE TERRACES.

THIS view gives an idea of the enormous size of the terraces formed by the siliceous deposits of the cascade of Otukapuarangi. The man shown in the photograph is a Maori chief named Parakia, the head of the tribe to whom the Rotomahana belongs. This Maori was a very aged man, and the whole of his body and limbs, from head to foot, was perfectly covered with a variety of tattooed designs. He did Mr. Mundy the honour to become his guide from Rotorua through the district. He was a man of fair stature, but the height of the terrace, as will be seen, was eighteen inches above his head. Underneath the overhanging edges of these terraces hang stalactites of various forms and tinted with different hues of pink, which may be gathered in plenty by a visitor who does not mind getting wet. The substance of the terraces is so hard as to require an axe for breaking off a specimen. It has the appearance of white china.

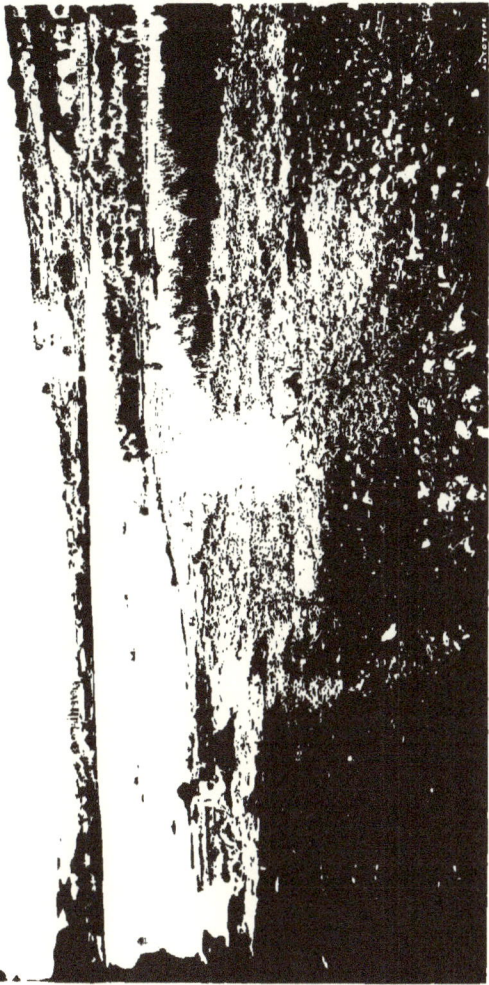

XII

LAKE ROTORUA, AND THE MAORI VILLAGE

OHINEMUTU.

WELVE or fifteen miles north of Rotomahana, towards the coast of the Bay of Plenty, is the Rotorua Lake with its beautiful scenery and picturesque surroundings, the third in size in this district, Lake Taupo being the largest, Tarawera the second. The name Rotorua means "Hole Lake," a lake lying in a circular excavation. With the exception of the southern bay, called Te Arikiroa, it has an almost circular form, with a diameter of about six miles, and a circumference of twenty miles. Almost in the precise centre of the lake is the peak-shaped island, Mokoia, rising about 400 feet above the level of the lake, with a pah or native fortress on the top, the scene of the Maori version of the legend of Hero and Leander, only that in this case it was the lady who took the swim. Her name was Ohinemoa. The circular form of the lake with the island in the middle, and the white steam clouds ascending along its western shore, might lead one to suppose that Rotorua had formerly been a volcanic crater; but in reality this lake, like all others of the district, was produced by the sinking of parts of the ground of the volcanic table land. The depth of the lake is comparatively small, perhaps at no place more than about five fathoms; it has numerous shallow sand-banks, and the shores also, with the exception of the north side, are flat and sandy. It is 1,043 feet above the level of the sea, and consequently of the same height with the Tarawera Lake; on the south-western side, the wood-covered Ngongotaha Mountain towers up to a height of 2,282 feet. This is the highest point of the range of hills encircling the lake. Among its numerous tributaries, the Puarenga River, emptying itself into the lake on the southern shore, near Whakarewarewa, is probably the most considerable; on the northern side the Ohau Creek forms the outlet of the lake into the Rotoiti, thus connecting the two lakes, which are separated only by a low and narrow isthmus.

The village of Ohinemutu, situated at the north-western extremity of the Rotorua Lake, is a native settlement, a famous old Maori pah, built on a peninsula. The dwellings of the chiefs are surrounded by enclosures of pole fences; and the Whares and Whare-punis, or dwellings and storehouses, some of which exhibit very fine specimens of the Maori order of architecture, are ornamented with grotesque wood carvings. The huts of the village are scattered over a considerable area on both sides of Ruapeka Bay, and on the slope of the Puke Roa Hill, rising to about 150 feet above the level of the lake. The place is noted far and near for its hot springs and its excellent warm baths (Waiariki). At Ohinemutu the Maori is seen living in the true Maori fashion. The camping-ground of His Royal Highness the Duke of Edinburgh, in December, 1867, with the Governor, Sir George Bowen, and his staff, was half-way up the hill to the right of this View.

Ruapeka Bay receives the water of the hot springs; here the water at the edge of the lake seethes and bubbles and steams in every direction. The principal spring is the Great Waikiti, at the southern side of the bay; its crater is but a few yards from the lake. Owing to the immense quantities of hot water here poured in, the water near the shore of the bay is warm, and forms an excellent bathing-place. By the swimmer approaching more or less near to the geyser, any degree of temperature can be obtained. The water of this geyser is perfectly clear. For some moments all is quiet in the large crater, only clouds of white vapour are seen ascending from it; then a powerful ebullition raises the water to a height of four or six feet, sometimes even to twelve. Little Waikiti, a few yards above, has a crater four or five feet wide, in which the water rises about every five minutes several feet high, sinking down again, during the intervals, to a depth of six or seven feet. The native women sitting beside the hole, as shown in the photograph, are cooking their potatoes and crayfish in it. The temperature of the water is 201° Fahr.

LAKE TAUPO.

DESCRIPTION of Lake Taupo has been given in the introductory chapter. This view is taken from near its northern outlet, on the banks of the Waikato River, and looks up the entire length of the lake to its southern shores. These are bordered by a succession of picturesque volcanic cones, behind which the Tongariro and Ruapahu rear their lofty heads. The Tongariro is clearly shown from the summit to the foot. With its regular conical form, bearing a still active crater, it rises majestically from the midst of a circular range shutting it in all around, and open only on the south-west side. The funnel-shaped crater at the summit of the cone can be distinctly seen, indeed, may almost be looked into, the west side of the crater being much lower than the east side. Thus the crater presents itself to view in the form of an ellipse, from which dense white steam clouds rise continually, sometimes enshrouding the whole peak, but at other times are driven far westward by the wind, affording a view of the blackened edges of the east side of the crater. The natives assert that the west side of the crater fell in at the time of the earthquake at Wellington, in 1855, and that at that time also a second crater to the north was active. Farther north, on the slope of the mountain, a brisk steaming solfatara is visible. The Tongariro in the summer months is clear of snow. The Ruapahu, shown to the left of Tongariro, has its summit frequently wrapped in dense clouds, below which the snow-fields of the peak are seen down to the height of 7,800 feet. At the base of these two colossal mountains, dark forests extend; in the foreground are mountains with sharp edges and deeply fissured precipices. Thus, at one glance, the effect of fire and water on the grandest scale may be taken into view.

From the south shore of Lake Taupo these two giants are not visible; but from the north end of the lake they are everywhere seen towering high above the lower mountain cones, which the natives call their wives and children. The names of these are, Pihanga, Kakaramea, Kuharua, Puke Kaikiore, and Rangitukua.

Pihanga, the most easterly of them, is the highest; it is estimated at 3,500 feet above the level of the sea. It is covered with wood, except only its topmost peak, which is cleft by a deep chasm, and displays from afar a crater open towards the north. The Kakaramea, the summit of which is of red colour, also probably bears a crater. Both these craters are supposed to be extinct; but the volcanic forces below have not yet been lulled into final repose, for on the northern declivity, and at the foot of the Kakaramea, the water steams, bubbles, and boils in more than a hundred places.

XIV

LAKE TAUPO, NORTH END, AT TAPUAEHARURU.

WITH ITS OUTLET INTO THE WAIKATO RIVER.

T the north end of Lake Taupo, the beautiful cone of Tauhara points out the region where the Waikato leaves the lake, even here a stream of great depth and force. This mighty river, it has been observed, taking its rise at the base of Tongariro and Ruapahu, flows through Lake Taupo, as the Rhine does through the Lake of Constance. At the base of Tauhara is a hot lake of a dark green colour. A stream from this runs across the plain, emptying itself into Lake Taupo, on the eastern side. Two miles farther on, down the course of the river, is the Otumaheke Valley. Here is a rivulet running quite hot the length of half a mile, and forming a waterfall close to the Waikato River. The ground, for miles hereabout, is boiled perfectly soft, thrown up into small mud cones, and pierced with many holes and crevices, discharging columns of steam ; some of these fumaroles, such as Karapiti, used to cast up water instead of discharging vapour, as they do at present. There is another region of hot springs at Orakeikorako, thirty or forty miles farther north; and again at Nihotikihore, the usual crossing place of the Waikato, on the road to the Rotomahana and Rotorua hot lakes. These terminate the volcanic district towards the coast of the Bay of Plenty.

XV

LAKE TAUPO, BOILING GEYSERS OF TOKANU.

GENERAL VIEW LOOKING NORTH-WEST.

HE south shore of Lake Taupo, at the head of the lake, is a flat delta formed by several mouths of the Upper Waikato. It is swampy to the edge of the lake, in some places, and covered with rushes, flax, and manuka scrub. Here are the remains of the village of Tokanu, which was destroyed by the rebel Te Kooti, like other Maori villages and pahs around the lake. In the middle of the low ground here is a series of geysers and mud-holes, occupying two square miles, from the small mountain Maunganamu to the mouth of the Tokanu Creek. This is really the head of the entire geyser system. The ground is paved in all directions with a white lime-like substance, covering spaces a hundred yards square. In this pavement are numerous boiling geysers of clear blue water, sometimes overflowing and adding by their deposit to the thickness of the crust; great care is needed in walking about the place. Several carved posts are set up here by the Maories to show where fatal disasters have occurred.

There are many baths here formed by the natives, converting this hot water to their use, alternately with the cold water of the Tokanu ; they also cook their food in the steam holes, which Mr. Mundy used for the same purpose during his stay in the neighbourhood. In the middle of this view is seen the Pirori Geyser, of which a separate illustration is offered on the next page. But nearer to the front is a warm creek, Te Atakoreke, with a temperature of 113 degrees, a favourite bathing-place of the natives. On the further side of the creek, three basins lie close together; Te Puia Nui (the Long Spout) is eight feet wide, and filled to the brim with gently bubbling water ; close to this are pools of greyish white mud, at a temperature of 188 degrees.

To the left of the present view is the site of Te Rapa, the native pah which was destroyed in May, 1846, by a landslip of the steaming ranges, and in which the great chief of Taupo, Tukino Te Heu Heu, perished with his six wives and fifty-four of his tribe. He was succeeded by his son, Iwikau Te Heu Heu, who now resides with his people at Waihi. In this vicinity is the beautiful double waterfall, 150 feet high, leaping out of the forest, and filling a rock-basin from which clouds of spray constantly rise, thence descending in a second cascade, to run into Lake Taupo. The lower fall had never been seen, even by the natives, till Mr. Mundy caused it to be cleared of the fuchsia and other trees which obstructed the view. He named it the Alfred Falls, in honour of the visit of his Royal Highness the Duke of Edinburgh to New Zealand.

MOUTH OF THE PIRORI GEYSER, AT TOKANU.

HE most powerful column of steam which is seen to ascend at Tokanu, and is visible far above the lake shore, belongs to the large geyser Pirori (a name that signifies fountain or eddy). It is situated on the left bank of the Tokanu Creek. A boiling hot-water column, six feet in diameter, always accompanied by a rapid development of steam, is whirled up at intervals of about eleven seconds, to a height which often varies. The geyser basin is eight feet wide, covered with a siliceous deposit resembling chalcedony; it is called the Jaws of Topohinga, and here the water is continually boiling. Mr. Mundy was informed by the officer commanding the district that he had seen a column of water a hundred feet high rising from it. The whole of the food for several hundred native and other troops was cooked in this geyser, when they were fighting the rebel Te Kooti in the neighbourhood.

A few years ago the Pirori discharged all its water, followed by quantities of boiling mud, stones, and steam, and several native women and children lost their lives in a neighbouring pah, on the banks of the Tokanu Creek, which is now "tapu" and deserted. No soundings have been taken since the eruption.

www.ingramcontent.com/pod-product-compliance
Lightning Source LLC
Chambersburg PA
CBHW021411090426
42742CB00009B/1107